Beyond The Twilight Zone

A journey through the depths of the ocean

By Delphine Mestdagh

Woodbridge Publishers
1200 Century Way, Thorpe Park,
Leeds, LS158ZA

First Edition

ISBN (Paperback): 978-1-916849-76-1
ISBN (Hardback): 978-1-916849-77-8

WOODBRIDGE
PUBLISHERS

Disclaimer

All animals in the book are identified by their scientific name, and the depths in which they move, while usually having a common name as well. It is also possible that different sources somewhat contradict the distribution of the animals discussed here. So, it is possible that you will read here that they live at a depth of 1000 m when, in reality, they have also been observed at a depth of 1250 m. I hope these discrepancies will be forgiven. Also, the boundaries of different habitats are not as clear as land boundaries are; again, different sources may contradict each other.

The selection of animals in this book was personal; there are, of course, countless more. All are equally beautiful, mysterious, and fascinating in their own ways. It was a tough call to make, but there was no other choice. Hopefully, all the animals not covered here will somehow come into the spotlight.

Acknowledgements

To Professor Ann Vanreusel and Doctor Lara Macheriotou of Ghent University, and to Professor Gudrun De Boeck of the University of Antwerp:

Because all three of you jumped into my submarine immediately and without hesitation. Because your passion for marine biology is contagious, but especially because you donated precious time in carefully proofreading and correcting my descriptions of the animals in this book: thank you very much!

And an additional thank you to Professor Ann for her inspiring foreword.

Preface, by Prof. Ann Vanreusel

A comparison made until today to explain the lack of knowledge about the deep sea explains how we know more about the surface of the moon than we do about the deep sea, even though it occupies nearly 70% of the Earth's surface, and represents more than 90% of the volume of our oceans.

However, our knowledge has recently increased very much thanks to the use of new technologies, such as robots and underwater drones, that allow larger surface areas to be observed or mapped over longer periods of time. The use of these new technologies equipped with high-resolution cameras also allows us to illustrate biodiversity through images that allow not only scientists but also the wider public to become acquainted with the wealth of deep-sea life.

A striking example is the recent insight that an extensive coral reef is present at about 1000 m depth along the Atlantic margin of Europe, so effective measures to protect these reefs have only been taken in the last decade.

The deep sea is, without a doubt, the most extraordinary ecosystem on our planet: there is no sunlight, no plants or algae grow there, the pressure of the overlying water mass is enormous at these great depths, and it is usually cold with the exception of those places where a hot mush is thrust from the bottom. For most animals, these "hot" spots near hydrothermal vents are poisonous, but some highly adapted species just occur there in large numbers, such as the yeti crab.

Even then, the most amazing thing about the deep sea is its great wealth of species and life forms. And it is that which Delphine illustrates so beautifully here in this wonderful book. She introduces us to some of the most remarkable but, at the same time, least known species and gives a glimpse of how and where these species survive in special living conditions. Some life forms are almost inconceivable, such as the Macropinna or barreleye fish with transparent skull, or the anglerfish, where the female carries her mate with her as a fused parasite and, thus, has a guarantee of sperm in her lonely life cycle. Also, the gulper eel, with its huge mouth and voracious way of life, outshines all the 'mythical fantasy creatures'. I hope that, like me, many people will become fascinated by the deep sea through this book and look with wonder at these extraordinary animals.

I further hope that with this book, the respect for nature and our planet will only increase, as for those ecosystems that are located far out of our sight, in almost inaccessible places. This is a book that makes us reflect on what nature is and what evolution can lead to. It is a wondrous journey through a literally dark environment where Delphine brings light with her beautiful paintings and clear texts.

Preface, by the Author

"Monster" according to the Oxford English Dictionary:

1) Originally: a mythical creature which is part animal and part human, or combines elements of two or more animal forms, and is frequently of great size and ferocious appearance.

   Later, more generally: any imaginary creature that is large, ugly, and frightening.

2) Something extraordinary or marvellous in its kind; an amazing event or occurrence; a prodigy, a marvel.

3) A malformed animal or plant; (*Medicine*) a fetus, neonate, or individual with a gross congenital malformation, usually of a degree incompatible with life.

4) A creature of huge size.

5) A person of repulsively unnatural character or exhibiting such extreme cruelty or wickedness as to appear inhuman; a monstrous example *of* evil, a vice, etc.

There are several meanings associated with the word "*monster.*"

The animals that live in the deep sea are also often called monsters. And I can't entirely blame the world for that. Granted, most animals living in the deep sea may not win beauty contests. They are not considered the fairest of them all, especially when photographed in an unflattering manner. Yet, I can't stand to hear it. How can I defend against such a strong prejudice?
I have always been prone to pity when it comes to animals. So, personally, when it relates to deep-sea animals, I like the second statement of the dictionary more: that they are extraordinary and marvellous. And honestly, I think they all possess a special beauty. Some because of the fairy-like light they emit, and others have beautiful colours; but above all, I have the greatest admiration for their way of life in those dark depths.

With this book, I would like to showcase these wonderful creatures in a new light. Really portray them at their best. I want to break their image. So that perhaps they will no longer be called monsters unless it is meant that they are marvellous.

Introduction

The ocean consists of five layers called "zones": the light zone, twilight zone, midnight zone, the abyssal zone and the hadal zone.

This book portrays animals that live below the light zone. This is the upper layer of the ocean and reaches about 150 meters deep. This zone is called the *epipelagic zone* and is home to as much as 90% of all marine life.

This makes sense because that's where the most food is found and the most currents that provide oxygen.

Photosynthesis can take place there, allowing various algae and seaweed to grow, as well as the phytoplankton, which is a huge food source.

Below 150 meters of that teeming light zone, it soon begins to dim. Colours disappear quickly; red becomes brown after only 6 meters, and in the twilight zone, brown is just a shadow. This is where much research ends, knowledge becomes fuzzy, and the number of mysteries begins to grow. Research becomes really difficult, especially when leaving the midnight zone! Although the recent unmanned submarines are technological marvels, it is difficult to traverse the vast ocean floor in search of living creatures.

Of course, this is not to say that there are no more animals to be found at those extreme depths! There are still so many, many I personally have yet to discover, and others have even managed to hide from the keen eye of science.

But who knows what the future holds? Perhaps there are even readers of this book who will later discover new species in the deep sea!

It may not be the vibrant, lively habitat we are used to in the light zone, but with a big spotlight, it is at least just as fascinating, exciting and colourful!

Searchlights on, and let the journey to the deep sea begin!

Chapter One

The Mesopelagic Zone

Also known as the Twilight Zone: 150-1000 m

In the twilight zone, some light still filters through, but too little for photosynthesis. Food becomes scarce, temperature drops rapidly to 4-5°C, water pressure increases, and conversely, the amount of dissolved oxygen decreases sharply. How has nature taken care of the animals that live here? What miraculous adaptations have they received? You will find out in this first chapter where some of the incredible inhabitants of the twilight zone are introduced.

***Tetronarce californica* (0-900 m)**

"Pacific stingray"

The first animal in this book is shocking. And quite literally, since it has two electrical organs on its head and can generate 50V. He uses this power to hunt or as self-defence; even humans can become unconscious from his shock! He has only small eyes, which he does not even need. The stingray lives in the twilight zone but still prefers the pitch-dark night to hunt.

During the day, it buries itself under a layer of sand; it does this with flapping movements of its wings. The wings are quite round, which is why they are also called the "disk."

Although it rests during the day, even then, fish such as herring and halibut have to watch out. The stingray senses their presence, shoots out of its hiding place, wraps its prey between its wings, gives a violent shock and devours the fish in the blink of an eye. And that too in less than a minute! Hence, its nickname: torpedo ray.

At night, it no longer hunts from ambush but actively pursues its prey. All this you would not expect when you see him floating at ease. He then looks so much like a flat frying pan that you would think he is not capable of so much. Appearances can often be deceptive and the *Tetronarce californica* is living proof of that!

Argyropelecus hemigymnus (100-700 m)

"hatchetfish"

This semi-transparent little fish grows to only 5 cm in size.

The animal's name refers to the shape of its body which, with a little imagination, reminds one of a hatchet. The scientific name also refers to it with argyros, meaning silver and pelekys, meaning hatchet. So, a silver hatchet...

They possess light organs on the underside of the body, the intensity of which they can regulate according to their needs. Their shadow is thus hidden from predators below them. This ability is called "backlighting," and many animals in the twilight zone take advantage of it. For the hatchetfish, this is very important because he himself is always looking up and cannot know what dangers are lurking below him. The eyes of the hatchetfish are very well developed and have a large retina, working like telescopes.

The *Argyropelecus* is really lightweight, weighing about 0.8 grams. They also possess a good swim bladder, which makes them appear to float. This weightlessness helps them in their daily migrations. At night, when they go up to chase their food which consists of small animals undertaking the same migration. In the morning, they migrate back to the safety of the twilight zone before dawn.

Chimaera monstrosa (300-600 m)

"dragonfish"

Dragonfish are most closely related to the shark, but that branch of the family tree split 400 million years ago. Thus, the dragonfish is one of the oldest fish species in the world.

It is a peculiar fish that has a few nicknames, with "ghostfish" being one of them. Their large eyes have a reflective coating and appear to glow in the dark. The pale, scaleless skin and the lines and dots running across the body make them look like skulls. Those lines are lateral line organs registering every movement in the water. The dots are electroreceptors that detect all electrical contractions up to and including the beating of a heart. This is useful for finding prey hidden under the sand or among the rocks.

They are also called "ratfish" because of their long, thin tail. The wing-like pectoral fins, in turn, are reminiscent of those of the ray. With these, they glide through the water and save precious energy. They possess a poisonous spine on their dorsal fin which breaks off after each use and a new one consequently grows.

Indeed, the chimaera appears to be an amalgamation of multiple animals, mirroring the mythological creature it's named after, which combines a lion, a goat and a snake. Interestingly, that legendary beast also had the ability to spew fire.

***Nautilius pompilius* (100-700 m)**

"nautilus"

The nautilus is a cephalopod, which literally translates to "head foot" in Greek, just like the octopus. They are called so because their tentacles are attached to their heads. Pretty simple. However, that's where the similarities end. The nautilus does not have eight tentacles like the octopus. Instead, it has 90, with which it searches the sea floor for goodies. Neither does it possess suction cups, nor does it squirt ink.

On the other hand, it is the only cephalopod to have a beautiful shell with a red tiger pattern, having been born with it, making him the perfect mini-version of his parents. The shell then has four chambers, with the creature living in the outer one. Its pearly home can be closed with a flap that resembles a cap. He lives in the same shell all his life and when the living room becomes too cramped, he makes a new one, closes the previous and never returns to it. All chambers are filled with water and air. The air-water ratio can be controlled by the animal, allowing it to rise and fall. It cannot swim forward, but with the help of jet propulsion, it can swim backwards with a rocking motion. At a full speed of no more than 1km/hour, it sometimes collides with a rock wall. This is why many nautilus shells found have cracks or dents. Still, that doesn't stop shell collectors. The *Nautilius pompilius* has been bobbing around in the deep sea for 500 million years. But in recent decades, its existence has been seriously threatened by this collecting frenzy.

Latimeria chalumnae (150-700 m)

"coelacanth"

The coelacanth is a large fish, up to 2 meters long. Its body is covered with large, hard scales that protect it like armour. It is strange that such a large, armoured fish was first seen only in 1938. Before that, it was known only as a fossil. The oldest fossils date back more than 400 million years. It was widely believed that the coelacanth became extinct 65 million years ago, at the end of the Cretaceous. It is a unique fish, in part because of its way of swimming. First, it moves its left pectoral fin and its right ventral fin, then its right pectoral fin and left ventral fin. In fact, it's exactly how a four-footed animal walks! Moreover, the fins are made up of real bones and look a bit like our limbs. From this, we can understand that the coelacanth is a link in the evolution between common fishes and the first animals that landed long ago. From these, amphibians, birds and mammals evolved!

All this time, the coelacanth has inhabited caves on steep, rocky slopes. There, it rests during the day and goes for a quiet walk at night. Then, it hunts its food at leisure. It himself is hunted by sharks.

Despite existing for more than 400 million years , its armour, finding safe haven in caves and having a limited number of predators, unfortunately, only 2 populations are known worldwide, making the coelacanth a critically endangered species.

Eptatetrus stoutii (10-1000 m)

"hagfish"

The hagfish is another living fossil: it remained unchanged for 330 million years. It is best known (or infamous) for its slime production. This gave them the nickname of "slime eel." When harassed, it releases a protein from the pores on the sides of its body, which becomes a terrible slimy substance when in contact with seawater that is not only difficult to remove from fishing gear but also ends up clogging gills.

Hunting fish, therefore, usually only venture one bite and immediately release this unsavoury lump before fleeing on their own. To extricate itself from its own "mucus dress," the hagfish ties itself into a knot and pulls its long, eel-like body through it. Then, it is free again to squirm wherever it wants.

Hagfish are the ocean's scavengers and feed almost exclusively on carrion, such as a dead whale that has sunk to the bottom.

For this, the hagfish has no jaws or real teeth but two pairs of rasps on its tongue with which it rips off flesh. With much wriggling, the hagfish works its way inside. But the question is, how does it breathe when half its body is in its dinner? This is easy for *Eptatetrus stoutii* as it breathes through its skin.

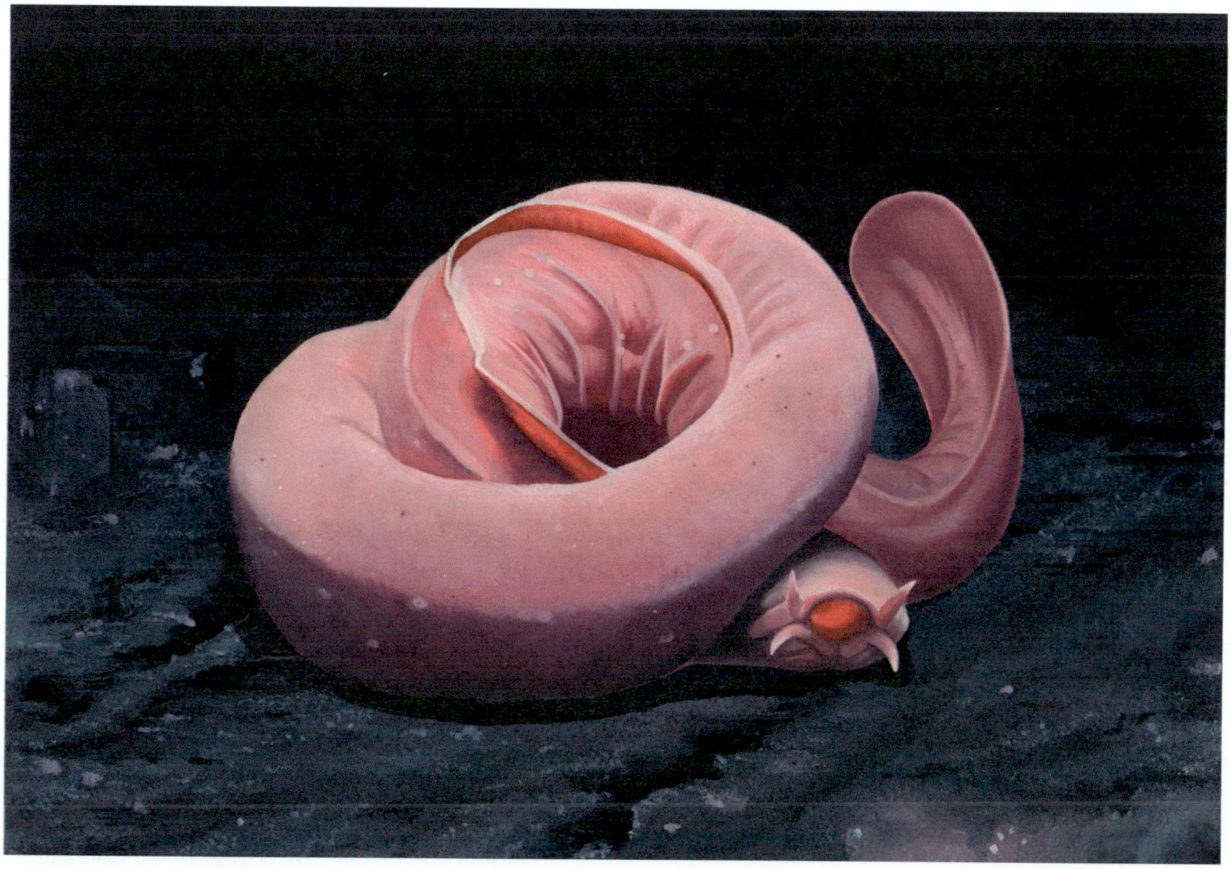

Regalecus glesne (20-1000 m)

"giant oarfish"

The oarfish is one of the enduring mysteries of the deep sea. Little is known about their lifestyle because only a few specimens have been seen alive. When they come to the surface, they are usually injured, dying, or already dead.

It has been claimed that they wash ashore due to tremors in the depths, and it was long believed that seeing an oarfish was a harbinger of an earthquake or tsunami. That turned out to be a fable with only a grain of truth, for the oarfish is not a very powerful swimmer and cannot withstand strong currents caused by earthquakes, among other things.

The oarfish has also long been chronicled as a true deep-sea monster, having been labelled as a giant sea snake. It is indeed huge, with the largest having said to have been a total of 11 meters! However, the average length for an oarfish is between 3 and 8 meters. This makes *Regalecus glesne* the longest bony fish in the world.

Its appearance is barely monstrous but rather graceful and beautiful. It resembles a silvery ribbon and has a long, red dorsal fin. By waving its dorsal fin, it moves slowly forward. Most of the time, however, it hangs vertically in the water column, which is also how it is virtually invisible to enemies swimming below it and searching upward for a meal. You might assume from a monster that they also have large, tearing teeth. However, the oarfish has none at all! It is a filter feeder, so it lives primarily on plankton and small squid.

There are many question marks surrounding the mysterious oarfish, but we are now at least rid of the preconceptions.

Chauliodus sloani (500-1200 m)

"viperfish"

This peculiar fish is named after the viper, because of its fearsome teeth.

As you can also see from the picture, they are indeed very long, translucent and sharp as needles. They are so long that they don't even fit in its mouth. When the viperfish closes its jaws, the teeth go along its face and the tips of the longest ones come right up next to its eyes! He can't chew with them, of course. In fact, the viperfish does not chew at all; its teeth act like prison bars, preventing the prey from escaping before it is swallowed whole. But fear not, perhaps you can also deduce from the illustration that the viperfish is not that big: 35cm at most. Its prey then obviously does not consist of humans, but small animals like the shrimp also pictured.

The viperfish can go a long time without eating: at least 12 days. Then, when no shrimp pass its fearsome mouth, the viperfish moves on to larger prey. To do so, it is well provided by Mother Nature with a bonus unique to *Chauliodus*, namely a hinge between its skull and spine that allows it to open its mouth wide and swallow prey 65% of its own size! In turn, to hide from hungry fish, it has a double row of light organs. These provide backlighting, just like in the hatchetfish we saw earlier.

Macropinna microstoma (600-1000 m)

"barreleye"

Again, the animal's name betrays one of its special characteristics. This time, you would assume something must be going on with his eyes.

You might suspect that the eyes are in front and that the fish is hanging out, looking somewhat pitifully ahead. But nothing could be further from the truth: those are the nostrils. The eyes are the two large, green bulbs in the head! With those remarkable eyes, the barreleye can see well in the dark; even the trick of backlighting doesn't fool this fish!
And if that's not striking enough, the barreleye can see through its skull!

The transparent dome not only gives the fish a good view, but it also protects its marvelous eyes from the stings of one of his favourite meals: jellyfish. To get those, he only has a small mouth. So, the barreleye must be able to strike very precisely. For that, the eyes help him again. When he sees a snack swimming by above, the barreleye gets ready. It points its nose at the prey and turns its eyes like binoculars in the same direction! The distance is then estimated, and the barreleye shoots at it with deadly precision!

Greatness comes in all sizes: *Macropinna microstoma* measures only 15cm.

***Vampyroteuthis infernalis* (600-1200 m)**

"vampire squid"

This squid grows to a maximum of 30 cm, but its name suggests that it is a bloodthirsty monster. *Vampyroteuthis infernalis* literally means "vampire squid from hell".

The name is certainly consistent with its appearance: it is blood red and has spiny tentacles. When it floats weightlessly in the darkness, it is as if it is wearing a dark cape. The similarities to the terrifying Dracula are telling. Yet the vampire squid is not blood-lusting. On the contrary, it eats marine snow that it catches with two long threads set with fine hairs. Marine snow is formed by mainly dead plankton whose remains swirl to the depths as snow. The squid makes balls of it and eats them.

Furthermore, the V*ampyroteuthis* is calm, rather shy and somewhat fearful. When it feels threatened, it slaps its tentacles over its head and hides in its "cape", whose spines now stand out menacingly. This, of course, makes him very uninteresting as prey.

Due to the low oxygen content in the deep sea, the creature moves little and very slowly, but in an emergency, it can still get out very quickly by jet propulsion. In flight, it uses light organs at the end of its tentacles to confuse its attacker. Two larger light-sensitive sensors on top of its head act as ghostly mock eyes, and although this squid does not squirt ink, it can still secrete a luminous ooze in case of panic. All this keeps other animals at bay, which is what the vampire squid likes best of all!

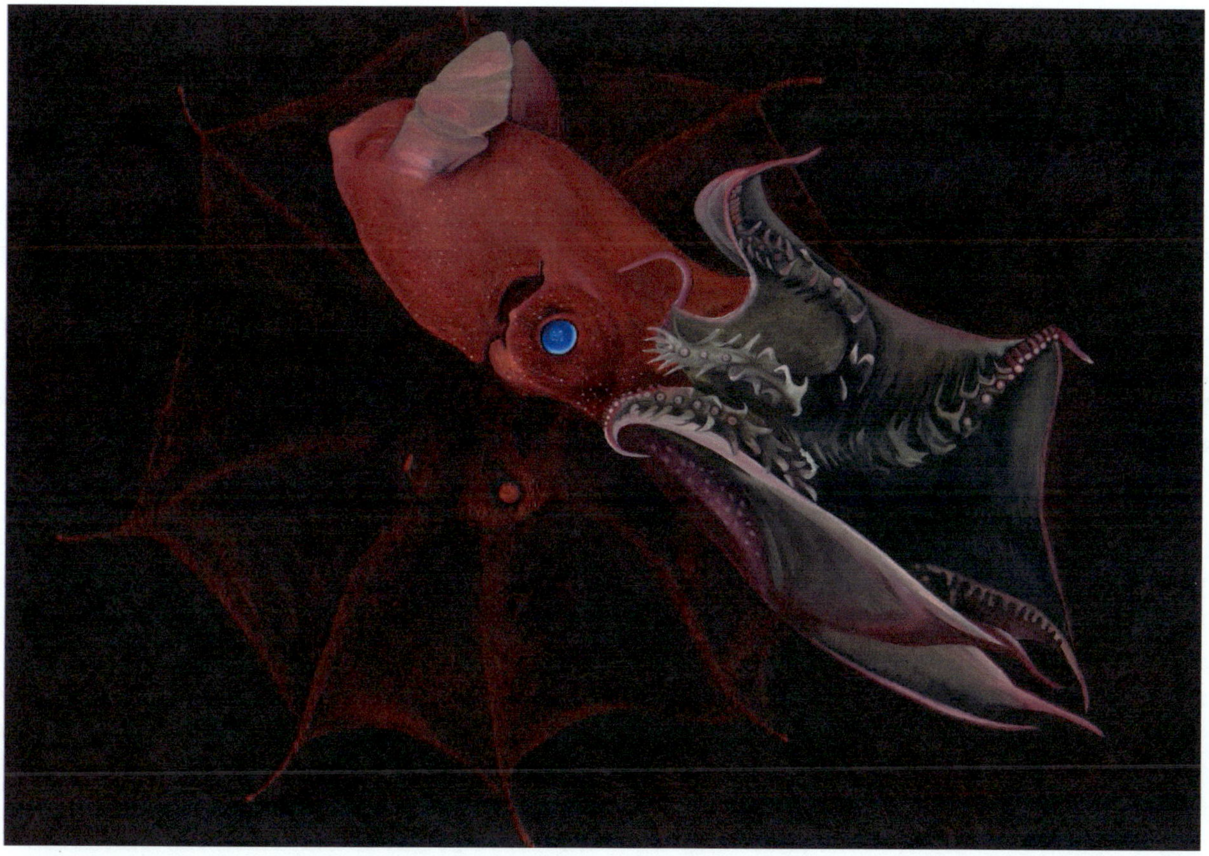

Psychrolutes marcidus (600-1200 m)

"blobfish"

The *Psychrolutes marcidus* is one of the most famous animals in this book.

Do you recognize it? No?

That's probably because I painted the fish as it is in its natural habitat: among the rocks some 1200 m deep off the coasts of Australia and Tasmania.

When he was first seen by human eyes, he was sadly dead. He had been accidentally caught in a dragnet and brought to the surface. With the release of tremendous pressure, his body turned into a hideous, misshapen blob. By this name and appearance, this wretch became known. Trawls are a great enemy of countless marine animals, not least the blobfish, which is threatened with extinction.

I, of course, prefer the live *Psychrolutes*. Did you know that this fish can live more than 100 years? And that the female lays 100,000 eggs in a "nest" on top of a piece of rock? The dedicated blobfish does not leave the nest until the pink pearly eggs break out. This waiting takes her literally no effort because these fish have very little muscle. And not much of a sturdy skeleton, either. This allows *Psychrolutes* to withstand the tremendous pressure. Meanwhile, they just eat whatever passes by their big mouths: crabs, shrimp and marine snow.

The blobfish has often been voted the ugliest animal in the world, which is quite sad.

One flip side of this pathetic coin is that its popularity has grown tremendously, and its protection is also getting some attention.

Grammatostomias flagellibarba (0-1500 m)

"dragonfish"

Seeing the illustration, you might think that the dragonfish is a typical deep-sea fish.

This little fish is also fairly small, barely 15cm. It can open its mouth very wide. It also possesses the typical long, needle-like teeth, the large eyes to catch maximum light, the light organs on the belly and a lure on a wire. Yet this animal is very special.

First, it is one of the darkest creatures on Earth. Darker even than the pitch-black night in which it lives. This is its camouflage, along with the rows of light organs on its belly. But there is something else very special about this dragonfish. As far as we know, this is the only fish in the deep sea that can perceive the red colour. All other deep-sea fish only see and/or produce blue light because blue light remains furthest visible. The dragonfish is not only able to see red, but it is the only fish that can also produce red light. Admittedly, this type of light dissolves after only 1.5 meters in the dark. However, that is enough for this hunter. When prey animals such as shrimp are within one and a half meters of the red light, the dragonfish can see them perfectly while he himself remains completely invisible.

So appearances are deceiving again. Despite its "ordinary" deep-sea appearance, *Grammatostomias flagellibarba* is unique because of Mother Nature's gifts.

Chapter Two

The Bathypelagic Zone

or Midnight Zone 1000-3000 m

The *mesopelagic* zone ends about 1000 meters deep, where the sparse twilight is dissolved into dense darkness. This is where the midnight zone, or *bathypelagic* zone, begins, extending to about a depth of 3000 meters. This is an extreme habitat, and so are the animals that live there! It is very difficult to study them; because of the immense pressure, it is difficult for people to reach them, and the animals cannot be collected alive for research without much difficulty. This is because their body tissues are mainly composed of fluids that have the same pressure as the water of their environment. When they are brought to the surface, they soon die.

Halitrephes maasi (1000-1500 m)

"firework jellyfish"

The firework jellyfish does not seem at all appropriate in a book about deep-sea animals. One would rather expect this jellyfish with its splashing colours to be found, for example, in warm, sunlit coral reefs where all creatures wear such colours.
Yet the *Halitrephes* feels perfectly at home in the deep sea. Its colours are invisible there because of the darkness, so it can live unseen and undisturbed.

This species is considered rare, as few people have had the honour of meeting one and witnessing its exploding colours.
As a result, unfortunately, little is known about it except that its bell-shaped body is about 10 cm in diameter, and its beautiful colours and pattern are caused by channels that run radially through its body. Those channels provide nutrient transport.
In this illustration, the *Halitrephes* is joined by a feather star. That is a type of free-swimming sea lily, another living fossil that peaked in the Carboniferous. Most sea lilies spend their lives anchored to the sea floor, but not the feather star. By waving its feathered arms, it moves forward.

The arms act as filters that retain edible particles. Because of the lack of strong currents in the deep sea, the feather star can swim freely without fear of damage with still a good chance of getting enough food.

***Architeuthis dux* (300-1500 m)**

"giant squid"

The giant squid is the second largest squid in the world after the colossal squid. Females grow just slightly larger than males, 13 and 10 meters, respectively. These deep-sea giants have the largest eyes known in the animal world: each eye is as big as a human's head! They have a beak very similar to that of a parrot. They have three hearts, blue blood, eight tentacles, plus two longer catching tentacles. Every tentacle is studded with suction cups, each of which has another circle of barbs. It has a perfectly streamlined body; when it holds all its arms together, it shoots through the water like a rocket. And that is just as fast forward as backwards!

His red colour keeps him well hidden in the dark, and in an emergency, he can also spray ink.

He needs this impressive equipment very badly despite *Architeuthis dux* only having a few enemies. But that one enemy does happen to be a real top hunter who prefers to see a giant squid on the menu. He dives to the depths especially for it and has made it his speciality to catch it.

That hunter is the sperm whale. It is not a deep-sea dweller, for it is a whale that breathes air like all other whales. But because of its exceptional ability to dive to extreme depths to catch an animal like the *Architeuthis dux* in the inky darkness, the sperm whale certainly deserves an honourable mention here.

***Caulophryne jordani* (100-1500 m)**

"anglerfish"

The anglerfish is a terrifying-looking fish with a large maw filled with sharp teeth. Her stomach is elastic, so she can swallow large prey in one go. She is dark brown to black, which makes her almost invisible in the dark all around. She is not very tall, barely 20 cm, but long fin rays envelop her like a net of filaments. She also possesses a rod on her forehead with a whitish, luminous tassel. With these, she lures curious, unsuspecting animals to her mouth. She is not an active hunter. She swims slowly and weightlessly in the slow current, waiting for her meal.

I use the words "her" and "she" because this is what the female anglerfish look like.

The male, who usually does not grow larger than a few centimeters is also seen in the illustration: the measly little appendage on the belly of Ms. Anglerfish.

He lives like a parasite from the moment he finds a mate. The male is initially free-living and has well-developed eyes and a strong sense of smell. This is to find a female in the dark. Once he has found her, he never lets her go. Quite literally, that is. He bites down and lives off her blood until, eventually, they grow together, and their bloodstream becomes one. Thus, the female always has a male with her when she lays eggs. Other than that, he does not cause her any trouble. In fact, even most of his organs die off.

Deepstaria enigmata (600-1750 m)

This jellyfish is one of the most peculiar creatures in the deep sea. It is a jellyfish, yes, which grows to one meter in size and is related to the common and widely known moon jelly. Only the *Deepstaria* looks more like a transparent plastic bag than a living creature.

The jellyfish was named after the submarine "*Deepstar*", with which the animal was first observed in 1960. *Enigmata* was added later on, which is pretty accurate since the animal is quite enigmatic.

What is certain is that the body of the *Deepstaria* acts as a balloon-shaped trap. It works very simply: a small fish or shrimp wanders through the opening inside the balloon, and the jellyfish closes the exit and consumes its meal. The channels you see in its translucent body are part of its digestive system.

The only animal that can survive in this "balloon of death" is a tiny Isopoda (a type of woodlouse).

So much for the certainties about this animal.

There is still much room for imagination, research and new riddles. Why does *Deepstaria* tolerate the tiny Isopoda? Why does Isopoda choose that very solitary existence? How come it is not killed by the stinging cells? These are also some of the many question marks surrounding this incredible animal.

Somniosus microcephalus (0-3000 m)

"Greenland shark"

The Greenland shark is a true record-breaker. First, they live in the cold waters from Canada to Norway and at 7 meters, they are the largest Arctic fish.

They grow slowly, only one cm per year. This is due to their slow metabolism. Actually, everything goes a little slower with this giant: they also swim slowly. This has made the shark an opportunistic hunter but also a scavenger. It eats fish, seabirds, seals and also drowned land animals such as reindeer. During the cold winters, these sharks often swim on the surface and have a chance to encounter such morsels.

They also age slowly. The Greenland shark is the longest-lived vertebrate in the world and is estimated to likely live for over 400 years! Think for a moment what times this shark has swum through! Humanity in Europe was busy with the likes of Mozart, Darwin and Napoleon.

Meanwhile, this giant wandered ignorantly into the depths. That is if it was not being caught by humans. Until 1960, they were fished en masse for their liver oil. Their meat is also poisonous unless smoked or double-cooked.

Moreover, their slow metabolism may account for their longevity, but also the fact that they only become sexually mature after 150 years! The eggs are incubated internally, and only ten young are born.

All this together makes *Somniosus microcephalus* very vulnerable, and it takes a long time for the population to recover.

Gigantura (500-3000 m)

"telescope fish"

When you see the name *Gigantura* mentioned and you look at the creature in the illustration, you might again think it is a giant, bloodthirsty monster. Let me reassure you. *Gigantura* is a compound name: "giganteus" is Latin and indeed translates to giant, whereas "uras" is Greek and means tail. So, the name refers to the long lower lobe of the caudal fin. The animal itself (including the tail) measures only 5 to 20 cm.

The long tail fin is one of the special features of this little animal. Furthermore, you could consider it a fairly typical deep-sea fish: it also has a mouth full of needle-shaped, sharp teeth that can slide in and out when catching and eating large prey. The *Gigantura* is silvery and, as a result, reflects the least amount of light of other bioluminescent creatures. To protect and camouflage itself, it usually swims vertically in the water column. This way, it is invisible to the hunting creatures below him. A survival trick we saw earlier with the oarfish. Its unusual forward-facing eyes have been adapted to this way of life.

Telescopic eyes are not uncommon in the deep sea, but usually, they are pointed upward, as in the hatchetfish and the barreleye. *Gigantura* thus appears to carry binoculars and is hence called telescope fish.

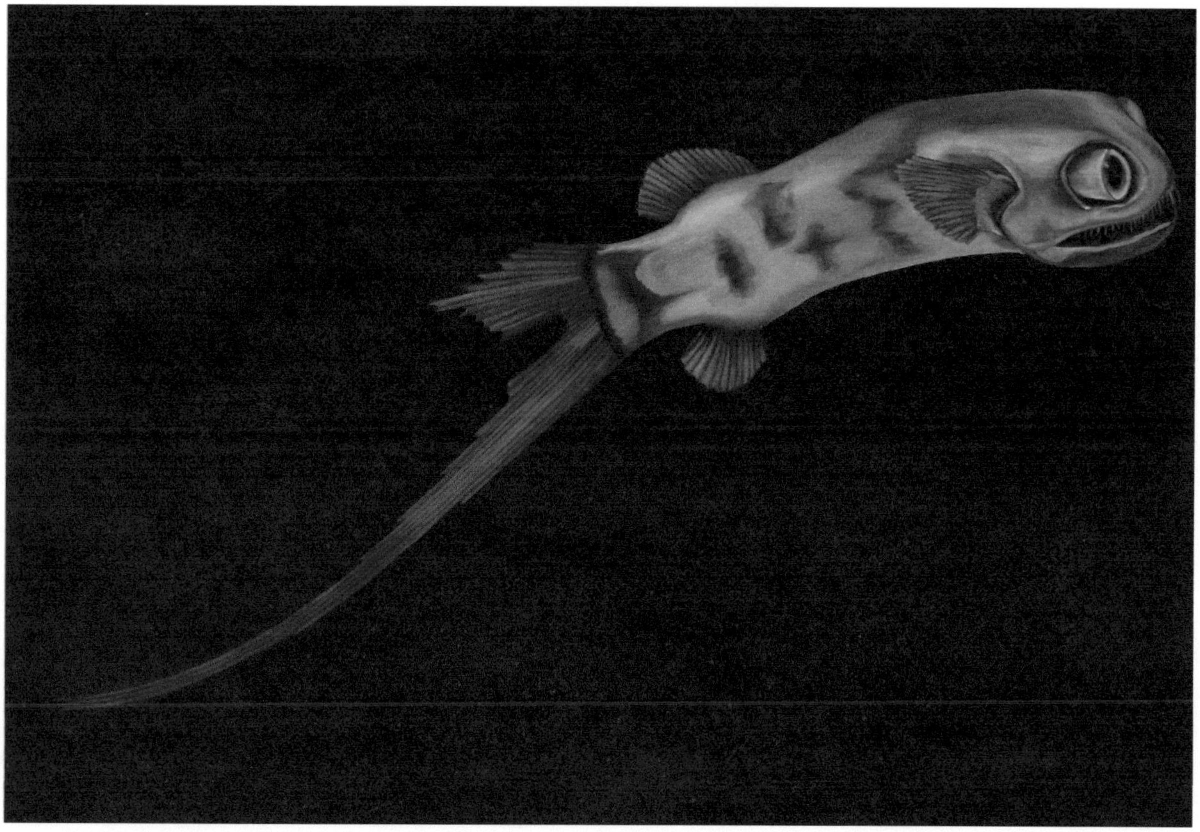

Siphonophorae (0-3000 m)

"siphonophores"

Siphonophorae is a collective name for a taxonomic group or order with up to 175 species. When talking about siphonophores, we do not refer to a single animal but an entire colony of many individuals called zooids. Individual zooids can't survive on their own and each fulfills its own task within the colony. Zooids are small, but all together, they can form huge colonies, with some even growing up to 45 meters long! Each zooid is a clone, descended from a single egg. Some of the clones are shaped like a polyp, while others are like a medusa. The uppermost polyp takes care of the buoyancy of the colony and others provide food, catching and killing prey with their thin waving tentacles that look so fairy-tale-like...

The first species of *siphonophorae* was described by Linnaeus in as early as 1758. This species is probably the most famous one of all siphonophores: the Portuguese man o' war. It is a species deadly to humans that lives on the surface of the ocean and floats on the waves.

However, many species live in the deep sea because they are built more fragile. They produce their own light through bioluminescence, giving them a magical appearance. The curtain of nettle cells then shines like fireworks in the black night.

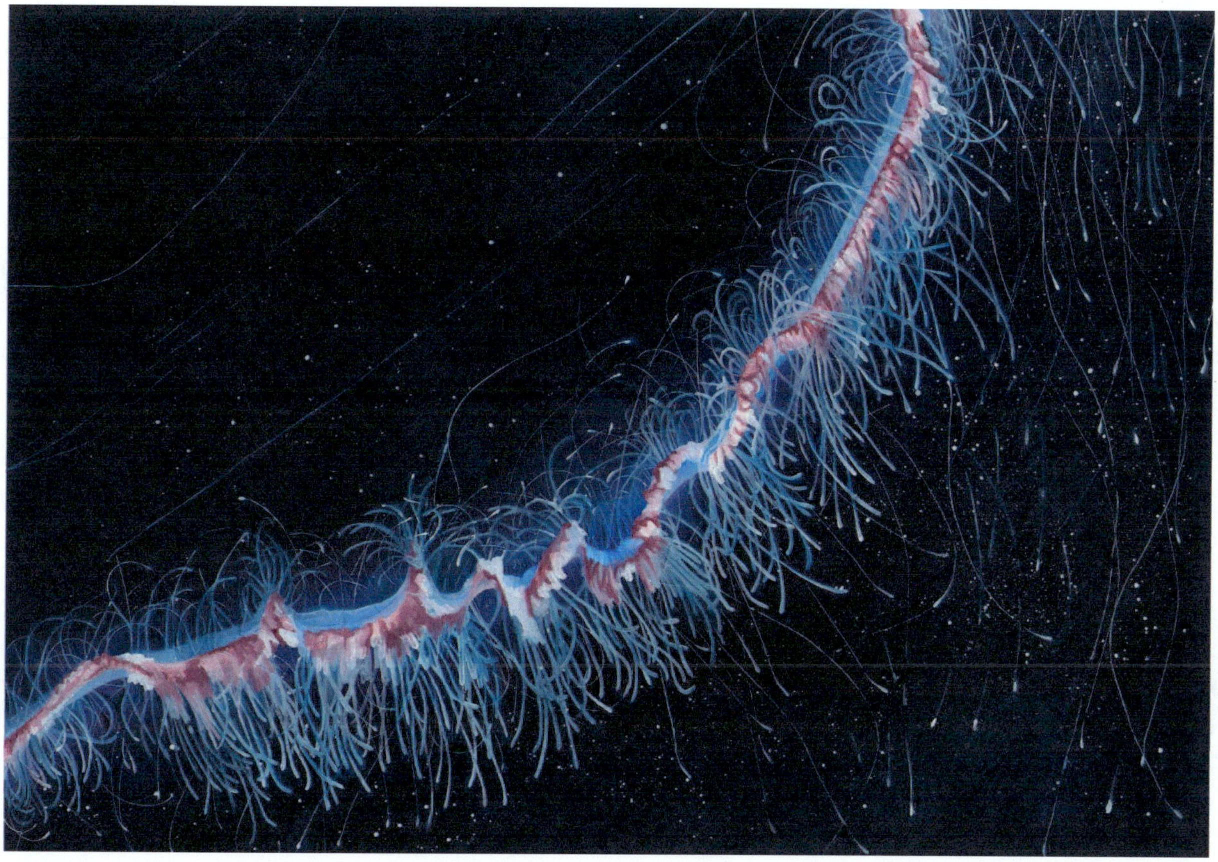

***Opisthoteuthis californiana* (130-3400 m)**

"dumbo octopus"

This has to be one of the cutest creatures in the deep sea. Discovered only in 1949, it immediately captured the hearts of scientists: it is often referred to as *Opisthoteuthis adorabilis* because of its high cuddliness factor. It is an octopus, but one that does not squirt ink. It has a small internal shell and two fins on its head. Those are the two "ears" that have earned it the nickname 'dumbo octopus'.

The animal swims exactly like a jellyfish, with those typical pulsating movements. Sometimes, they spread their tentacles and let themselves hang out in the currentless depths.

Another peculiarity on the part of the female is that she can lay eggs several times in her lifetime. This is very remarkable because after laying their egg clusters, most octopus mothers guard their unborn offspring without ever leaving the nest. They are so dedicated to their task that they do not even go hunting and, consequently, die of starvation. The small dumbo octopus is a bit looser with its offspring. Consequently, it takes nearly three years for the young to hatch and begin their lives.

***Stychiomedusa gigantea* (800-2000 m)**

This jellyfish is closely related to the *Deepstaria enigmata,* to which we were introduced a few pages ago. They both don't like to show themselves, something they have in common. The *Stygiomedusa gigantea* was first described in 1910 but has rarely been observed since, even though they are found in all of the world's oceans except the Arctic.

This time, though, *gigantea* is an appropriate name for the whole animal. The disc is one meter in diameter, and the tentacles are ten meters long! They have no stinging cells like most jellyfish we know. They are long, wide strips that wrap around small prey animals. These types of tentacles are called "oral arms," and the *Stygiomedusa* has four of them. There are two different types of oral arms: sometimes, the arms take the captured prey to the mouth, whereas the other kind has special mouthparts. These mouthparts are essentially small openings that eat the prey.

But forget these frightening details for a moment and note the mysterious beauty of this Guardian of the Underworld. The *Stygiomedusa gigantea* has a mesmerizing look, and it is hypnotic to follow its slow movements.

Kiwa tyleri (2500 m)

"yeti crab"

Earth's crust consists of nine large and six smaller tectonic plates. These plates have a thickness of about 100km and float, as it were, on layers of thick liquid. New plates are formed where the oceanic crust is still porous, where seawater penetrates the pores and heats up in the deeper layers. Here are the so-called hydrothermal vents. These are actually chimneys in the oceanic crust that release hot water that can be up to 300°C hot and rich in minerals.

Near those chimneys, where the water has already cooled a bit to a more pleasant 25°C, live the yeti crabs. On their hairy legs grow bacteria that serve as food and need the hot, mineral-rich water to survive and multiply. It is, therefore, a wonderful haven for the crabs.

The *Kiwa tyleri* lives in the northern and southern parts of the Scotia Plate near Antarctica. The females carry the eggs and then stay a little farther away from the chimneys. The young crabs also prefer it a little cooler. The males always stay on chimneys. They literally live there by heaps, with sometimes more than 700 crabs per square meter! It is, therefore, a good thing that the *Kiwa tyleri* can store its own food supply. Especially on the belly side, where they are richly hairy and can store their supply of bacteria to consume them later.

Chapter Three

The Abyssopelagic Zone

or Abyss 3000-6000 m

From 3000 meters, a new zone begins: the *abyssopelagic* zone or the part of the water column located at great depth above the abyssal plain. This zone goes up to 6000 meters deep. If the zones above it were already extreme, this one certainly is!

The abyssal plain is a mostly flat surface and the world's largest habitat. It has often been said that the surface of the moon has been better explored than the abyss. It is well known that the bottom consists mainly of clay and sludge deposits that, due to the pressure and lack of flow, appear very monotonous at first glance. The plains are intersected by submarine ridges and volcanic seamounts rising from the depths into the twilight zone or even higher. Here, you can travel for miles without seemingly encountering a single creature!

However, if you look closer, there is life, albeit sparse and sometimes, in the form of microscopic organisms.

Teuthidodrilus samae (3000 m)

"squidworm"

This is another animal discovered recently: only in 2007. In 2011, it was officially described as a type of bristle worm, so-called because of the two rows of twenty-five or more bristly paddles along the flanks of the small animal.

The scientists who described and named the little worm referred to the order of squids: *Teuthida*, because of the ten appendages that grow from the creature's head. The appendages are about 10 cm long, making them as long or longer than the transparent body. Hence, the comparison with the cephalopod lookalike. The appendages do not have suction cups as in the squid, but are still quite fascinating. Eight of the ten are actually extended gills and are thus used for breathing.

The other two (in the illustration, they are the curled-up ones) are sensory organs. This bristle worm also has two more feathery organs that act like a nose. The *Teuthidodrilus* swims slowly just above the sea floor. As it does so, it makes undulating movements with its numerous paddles. It is a beautiful, graceful dance of the ghostly creature.

Chondrocladia lyra (3300-3500 m)

"harp sponge"

While perhaps many of us thought the deep sea was a quiet, peaceful place on our Earth, by now, we will all know better.

Food is scarce there, and that can cause hostility. Some animals look somewhat sluggish, benign, sometimes even cute, and then they turn out to be dreaded hunters. Take the *Chondrocladia lyra*, for example. This doesn't even look animal-like at all with those strange, upright branches. Yet it is an animal that has been given the name harp sponge because of the bizarre structure of its body. And also because it is a sponge. Not the peaceful kind that filters plankton and bacteria from the sea by pumping water through its body. No, this is a sponge that eats meat.

Carnivorous sponges are a great exception, and there are only seven known species in the world! It catches small crustaceans with the fine hooks that line the vertical branches. Those hooks act like velcro where the nutritious shrimp get stuck. The harp structure is useful here because it grows 60 cm wide and 30-40 cm high and thus covers a fairly large area, the chances of snagging something tasty are higher.

Extreme conditions require extreme measures. This is proven by the carnivorous harp sponge.

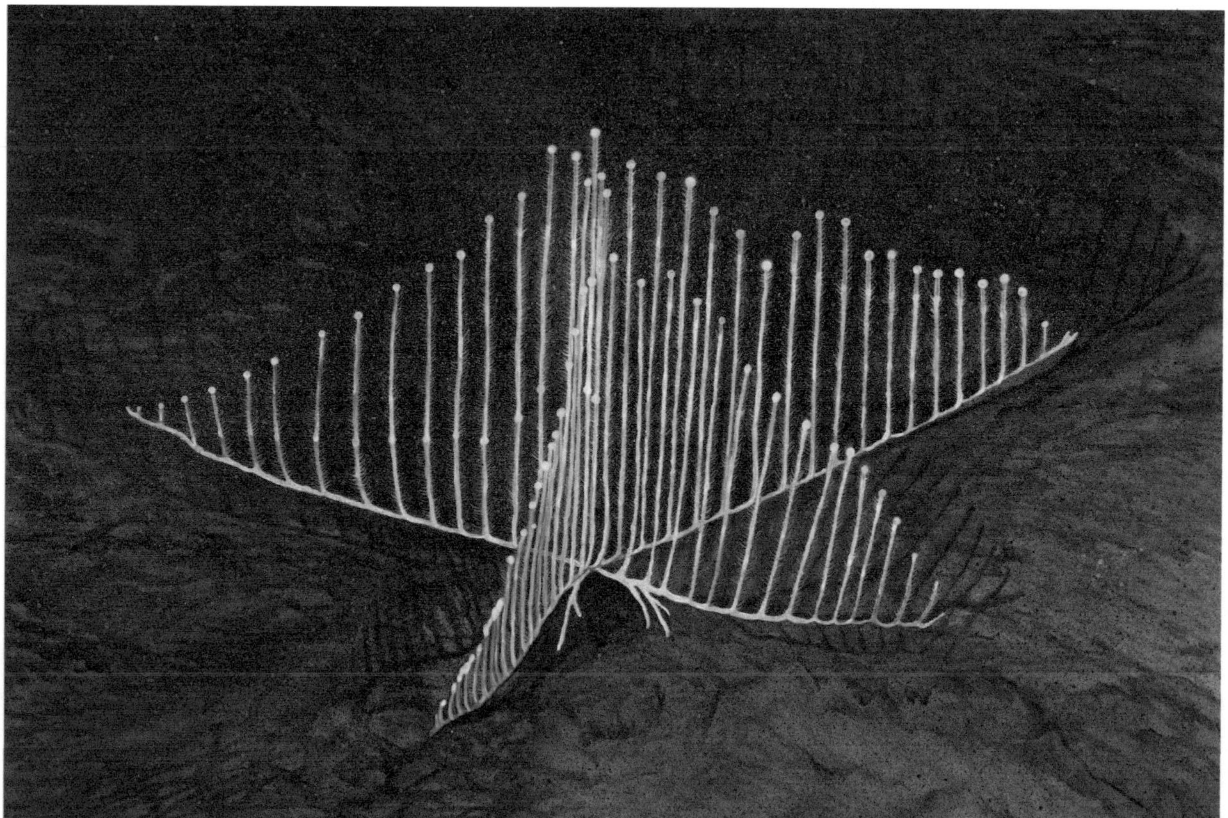

***Swima bombiviridis* (2700-3600 m)**

"green bomber worm"

There are about 10,000 known species of bristle worms in the ocean, and new ones are discovered regularly. Some are more bizarre than others. Once you open this "box of worms," another world of surprises awaits!

This book, unfortunately, has only two species, and they happen to be related.

This is *Swima bombiviridis*. "*Bombiviridis*" is a compound of two Latin words: "bombus" means buzzing, and the word bomb was derived from this. Green is the translation for "viridis." In other words, this little 3cm worm throws green bombs!

The little worm is also a powerful swimmer for its size and can move forward and backwards as fast. To do this, much like its cousin *Teuthidodrilus*, it uses brush-like paddles.

The green bomber is blind but senses when danger is imminent. This is when the bombs come in handy. They have eight of those, which are located just behind the head. They are actually pouches filled with a bioluminescent liquid that the worm can release at will. Lost bombs grow back.

When the bomb bursts open, a green glow is dispersed. In the resulting confusion, the little worm can quickly get away.

Duobrachium sparksae (3900 m)

This strange little balloon was first seen in 2015 through the eyes of an unmanned submarine. Until today, no human has seen it in the flesh. At the time, the remote-controlled submarine was not equipped to take the animal to the surface. So, there is only video footage on which the full official description was based; the first time such a thing has been done.

The ballooning animal is a comb jelly or *Ctenophora* and is recognized by its rows of cilia. When these *cilia* move in a spotlight, the light refracts, and all the colours of the rainbow become visible. The balloon itself measures about 6cm with two 30 cm long tentacles. This was measured using two laser beams placed at a fixed distance from each other.

The scientists noticed that the tentacles kept touching the rocky bottom and suspected that the comb jelly anchored itself to stay in position. The scientists hope to see this mysterious animal again soon to conduct further research. However, they also know it can sometimes take decades for it to reappear.

Spectrunculus grandis (800-4300 m)

First described as early as 1877, the *Spectrunculus* is one of the largest bony fishes of the midnight zone and the abyss at 1.27 meters. They have no scales but often have pale skin with white fins. Sometimes, they are also light brown to dark brown. Not a ray of sunlight breaks through in its habitat. However, its pale skin can reflect light from other bioluminescent creatures. That can be a disadvantage when you prefer to remain hidden in darkness. Perhaps that's why they live closer to the bottom.

That's also where they find their food. They mainly eat bristle worms, snails and carrion that swirls down. *Spectrunculus* belongs to the large order of *Ophidiiformes*, which includes many fishes that closely resemble eels and eel-like fish. The name *Ophidiidae* is derived from Ancient Greek, where "Ophis" means snake and points to the long, slender shape of their bodies. They are distinguished from eels by their ventral fins that are developed into sensory barbels at the base of the head.

Anoplogaster cornuta (500-5000 m)

"ogrefish or fangtooth"

This fish can also be found in the pitch-black void of the abyssal zone.

You could base some horror stories on the appearance of this little monster. It only grows to about 18cm, but with its long, pointed teeth and head that has scales resembling soldered steel plates, this deep-sea dweller does not look so lovable. Fortunately, that is not one of his worries, as this animal is equipped to survive at depths of up to 5,000 meters!

It is also often found in the twilight and midnight zones, but there, it often falls prey to tuna and marlin. In the abyssal zone, he is a lot safer! Because of its dark brown colour and special light-absorbing skin, it is totally invisible to predators with the so-called bioluminescent headlights. With its long fangs, it can spear prey as big as itself. To close its mouth, there are cavities in the upper jaw into which its teeth fit perfectly, like a knife in a sheath.

The *Anoplogaster* often swims in small schools, and for this purpose, it has a very well-developed lateral line system. In this fish, there is an open groove covered with hard scales. Underneath are sensory organs that connect to the brain. This allows each animal to sense where its comrades are.

***Enypniastes eximia* (500-7000 m)**

"Spanish dancer"

When mentioning a sea cucumber, we usually imagine a somewhat strange creature whose shape and colours remind us of a large caterpillar. Normally, they crawl rather cumbersomely across the sea floor. Yet the *Enypniastes* is also a sea cucumber, but one that breaks all the rules.

It swims up to 1000 m above the bottom and touches it only to eat. To do this, it has a circle of tentacles around its mouth. At the ends are small, flexible grippers with which it picks up sediment from the sea floor and puts it into its mouth. The sediment contains organic particles in the form of flakes of sea snow that have drifted down. In the illustration, you can see the gut through the transparent body. The gut filters the food particles out of the sediment, after which the remaining purified sediment is discharged back. So this little animal is also an important cleaner. As it scavenges its meal, it naturally churns up sediment. This benefits other animals living in the sediment.

It got its nickname "Spanish dancer" because of the way it swims. In doing so, he repeatedly bends forward and backwards with waving tentacles. With a little imagination, this does indeed look like a dance. However, it is not necessarily found only in Spanish seas but all over the world.

***Eurypharynx pelecanoides* (500-7600 m)**

"pelican eel/gulper eel"

It is impossible to ignore the remarkable names of the *Eurypharynx pelecanoides*. Clearly, they refer to its mouth. More on that in a moment but for now, let's talk about the rest of its body, which is equally fascinating!

The gulper eel is about half a meter to a maximum of 1 meter long, and has very small teeth and small, black beady eyes. They swim with wavy movements of their long tails. One big peculiarity is just in that tail. Most of it is a red thread, which ends white with a pink light at the tip. To hunt, it curls its flexible tail above its head (and here we inevitably come to the beak) until it reaches the wide-open mouth. The inside is very dark, velvety black and completely invisible to all eyes in the deep sea.

When prey comes within range, like a pelican, he takes a big scoop. Now he looks more like a mini whale with a throat pouch like a balloon full of water and hopefully some goodies to eat. The whole thing is swallowed, the water leaves the body through the gills and the pelican eel meanders on its way.

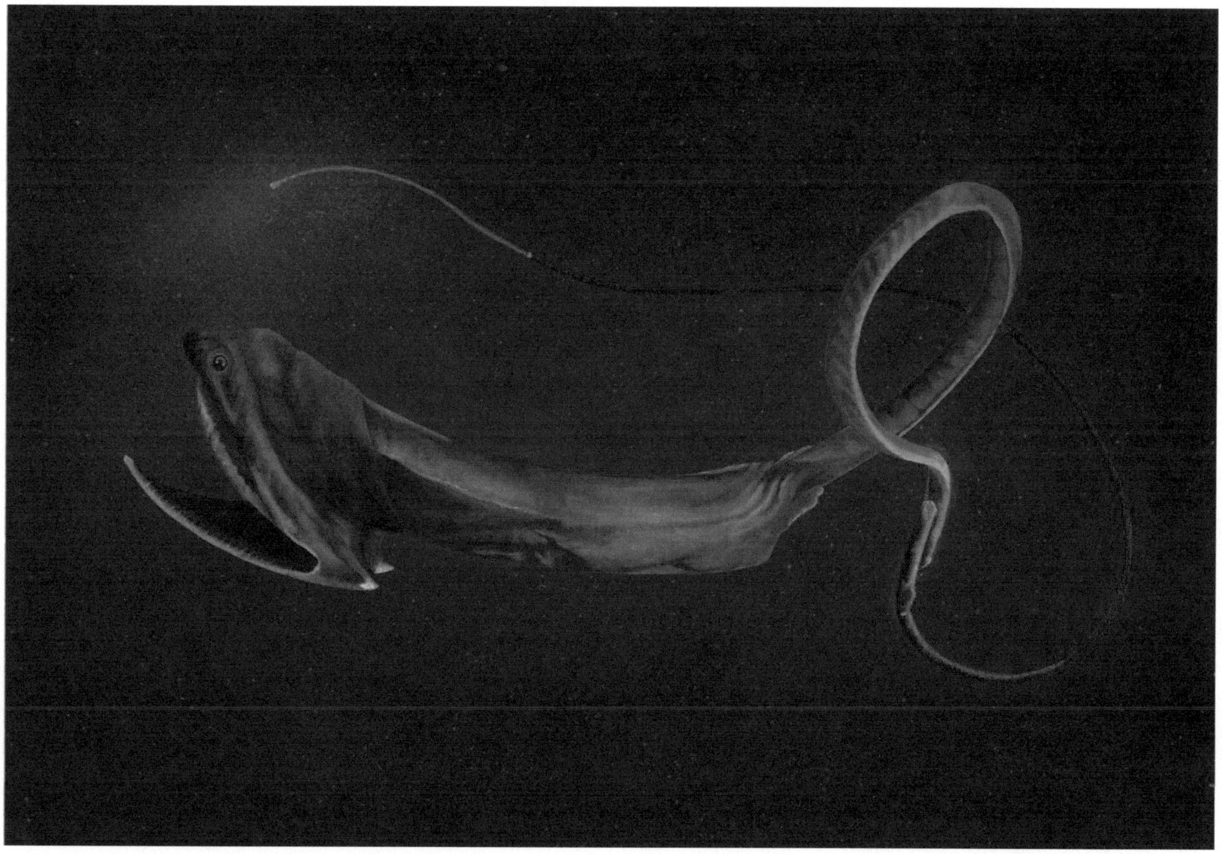

Chapter Four

The Hadal Zone

or Trenches 6000-11000 m

The abyss is the largest part of the ocean, yet there are places that are even deeper. These are the deep-sea trenches or *hadal* zone, named after Hades, the Greek god of the underworld. These are places so unearthly that even astronomers organize their field of activity there. For organisms that can survive in these extreme conditions might also exist on other planets...

Pseudoliparis swirei (7000-8000 m)

"Mariana snailfish"

I can already imagine what you are thinking upon seeing the illustration. Perhaps you horror a little at the thought of ghostly tadpoles wandering in the underworld; hungry, limp skin over bone, grinning in the dark... And in the hadal zone, you never know what you might encounter.

But rest assured, it is neither a ghost nor a tadpole. It is a full-sized fish of nearly 30 cm, one of the most incredible in this entire book.

It is (for now) the record holder for the deepest living fish in the world! This fish withstands pressure 700 to 800 times greater than at the surface! The *Pseudoliparis* has no natural predators, so it is also bombed as the top predatory fish of the hadal depths, even though it catches mostly small crustaceans.

I also wrote that this fish is a record holder, but only for the time being, as there are contenders for the honorary title. Only the sightings are still somewhat doubtful, or the species has not yet been officially described.

So you read it right! There is still a lot to discover here. But for now, the incredible *Pseudoliparis swirei* still sits on the throne.

Scotoplanes globosa (1000-10,000 m)

"sea pig"

We have come to the end of this book. This is the last animal I would like to highlight. And what an animal!

These have been spotted at a depth of nearly 10,000 meters, and they can grow to a maximum of 17 cm. They are piglet-pink and walk on the sea floor with short, stubby legs while rooting in the mud with their snouts. Sounds familiar? Well you're right, it's a sea pig, of course!

Actually, it is a kind of sea cucumber, and it doesn't really walk on legs either, but on so-called tube feet. These are similar to the feet of starfish and other species of sea cucumber. Only the feet of sea pigs are unusually long, ideal for walking on soft mud.

Also different from the pigs we already know, their snouts are tentacles arranged around the mouth and used to sift decomposing bits of algae and animal waste from the sediment. The antennas at the top are also a kind of tube feet (called papillae). They do not serve to walk upside down, however, but to detect food. When they have found something large, for example, a whale carcass, the sea pigs gather in huge flocks for a veritable feast.

"Accuse not nature, she hath done her part, do thou but thine."

(Paradise Lost, by John Milton)

Delphine Mestdagh
Caesar Gezellestraat 43
8560 Moorsele
Belgium
BE 0791.803.080

Follow @vinnies_traveljournal on Instagram

Artist Member of the Deep-Sea Biology Society

www.dsbsoc.org

The mission statement of the DSBS:

The advancement of education for the public benefit in deep-sea biology by the operation of a scientific society to promote the exchange of information, the promotion and dissemination of deep-sea research, to foster the next generation of deep-sea biologists and to promote demographic diversity in the study of deep-sea biology.

DEEP-SEA BIOLOGY SOCIETY

Made in the USA
Monee, IL
05 November 2026

67387141R00050